CIDER MAKING FROM YOUR GARDEN

by Charlie Henley

Published by Forest Fountain Books,
forest.fountain@btinternet.com

First published as a Kindle e-book, 2011

Revised as a Paperback, 2017

Obtaining this book
Both the paperback and the much cheaper Kindle e-book
edition can be bought on-line from Amazon.

Prospective retailers can discuss options via the e-mail at the
top of this page (answered weekly).

CONTENTS

All you need for cider: fruit press, food processor &
fermentation barrel .... plus an apple tree or two

1. MAKE EVERY APPLE A WANTED APPLE!

This booklet shows how to make good cider from just one or two apple trees in an ordinary garden – with minimal equipment or knowledge.

For North American readers, we're talking about fermented, alcoholic, apple cider – called 'hard cider' over there but 'cider' here in England.

Our cider-making began when we moved to a house with three apple trees in the garden. There was one very fruitful tree, one very small tree, and one tree whose yield varied every year. However many apples we ate or stewed for the freezer, more still rained down to litter our lawn. So for 16 years now we have made cider from our spare apples. Now every apple is a wanted apple.

Usually we press our first barrel from windfalls in early September and our last barrel at the end of October.

Typically we produce 9 – 16 gallons of cider per harvest (55 – 72 litres), though unusual harvests have taken this as high as 23 gallons (105 litres) and as low as one gallon (4.55 litres). By gallon, please note, we mean the traditional British Imperial gallon which equals 1.2 US liquid gallons.

Our methods are as low tech and simple as they come! No chemicals added whatsoever. Not much scientific understanding either! Nor that much spending on equipment. Certainly there are other ways you could do it, but follow our simple approach and you will get a reasonable result.

2. WHAT DO YOU NEED TO USE THIS METHOD?

Apples

In terms of type, these are whatever you've got on your own apple trees - and mixing apples from different types of trees is a common cider-making practice. True, there are some special varieties of apples traditionally used for cider in Britain. But any apple will produce a result. Once your first cider is ready, you'll find out if your particular apples are skewing it too sweet or too dry and you can adjust the process at your next harvest. You can make cider anywhere that sufficient apples grow, be it Scotland or Tasmania. We're in north Yorkshire, well north of the major cider region in south-west England.

Enough apples to make this worthwhile

We expect at least one gallon of cider (4.55 litres or 1.2 US liquid gallons) from 18 or 19 lbs (8 - 8.5 kilos) of our uncut apples. (So, very roughly, 2 kilos of apples yield a litre of cider.) But by late October, when apples are juicier, we have sometimes got a gallon from only 14 lbs of apples (6.35 kilos). Apple trees vary in juiciness and you'll need to measure your own results. But this gives you a rough rule of thumb for calculating how much your trees may give you.

In early autumn you could collect early windfalls in a bucket under the tree, then take out bathroom scales and weigh how much you are accruing.

Or, by early July, small apples will be easily visible and you could count or estimate the total number on a tree. For our apple trees, it can take 50 - 60 apples to produce one gallon of cider (4.55 litres or 1.2 US liquid gallons).

Unless your trees produce extraordinarily small apples, this offers a crude predictor for eventual cider volume. (But when actually gathering apples for an estimate, always weigh, not count them, for apples can vary so much in weight.)

Key equipment

Fruit press

The simplest type is a cylinder into which you place a muslin bag of mashed apple and screw down a tight-fitting block to compress the mashed apple. Currently we use the smallest Vigo press, Vigo's stainless steel '5.3 litre' model. It's shown in this photo, screwed down.

New fruit presses can be expensive – they're the only expensive item here - so second-hand options are well worth investigating. Larger fruit presses than ours can considerably reduce the time required. But they're even more expensive. See Chapter 6 for sourcing fruit presses like our own.

See '**Saving Money on the Fruit Press?**' in Chapter 5 for options like making your own fruit press or getting by without one. One such option we will challenge right here: kitchen fruit juice machines. We haven't seen a model which was not far, far too slow for the amount of apples which many gardens yield.

Remember that the early bird catches the worm. Each September, as the apples start raining on all those lawns, there's a run on every sort of cider-making equipment.

Ordinary kitchen electric food-processor
In traditional English commercial cider-making, before being placed in a large fruit press, the apples were first broken up. Often this was done by huge roller millstones turned by horses. In our method an electric food-processor is used instead for this purpose.

There are three differences between this and the small apple crusher machines which are commercially advertised to home cider-makers:
- The finer mashing by the food-processor results in cloudy brown apple juice. But, don't worry, once your cider is ready, it will be completely clear.
- This finer mashing produces appreciably more juice per apple.
- If you've got an electric food-processor already, there's nothing extra to buy.

A household electric food-processor is great for breaking up / pulping / mashing the apples before pressing. But many home cider-making recipes have not cottoned on to this. Surprisingly often you see recipes telling you to pound the apples with heavy timbers – laborious in the extreme!

Double muslin bag
This holds the apple, mashed by the food-processor, and goes inside the fruit press. We use two, one inside the other, in case of breakage.

Lidded plastic 'fermentation barrel'
Like the muslin bags, you can get these from a homebrew shop (homebrew shops can supply everything mentioned). These barrels are simply barrels with lids – see photo on page 3. We use 2 gallon, 3 gallon, and 5 gallon sizes (roughly 9 litres, 13.5 litres and 23 litres respectively). But you need only one barrel for your first 'fermentation', a process which we'll explain at the right moment.

Since exposure to air can seriously harm cider, we fit the lid with an airlock (see photo overleaf) so the lid can be closed completely tight once fermentation slows. Homebrew suppliers sell both airlocks and special rubber washers for thus attaching airlocks. You make a neat hole in the lid and one side of the washer grips the lid on the outside and the other on the inside. See '**How to fit an airlock'**, page 54.

If seeking to avoid paying for airlocks, on You Tube you'll find videos of various homemade airlock devices.

Specialised cider fermenting vessels, which protect against air, are also available. But they are often hard to obtain in a small enough size.

Fermentation barrel with airlock

We'd advise against pressing more than 5 gallons (23 litres) at once – too tiring. For your very first fermentation, aim for 2 gallons (9 litres) of apple juice. Go for larger volumes only once you know the work entailed in pressing.

Demijohns with airlocks
These are 1 gallon glass vessels for storing cider, post-fermentation. They have a tight cork with an airlock so that CO_2, from any continuing fermentation, can escape but air cannot get in and harm the cider. The demijohns pictured on the page opposite have an older type of airlock to that in the fermentation barrel above. The latter includes a little cap to exclude flies, often attracted to fermenting liquid. It can be seen close-up on page 25.

Demi-johns with an older type of airlock

A well-chosen yeast

Yeast is needed to trigger fermentation. Our method differs from the most traditional cider-making, where natural yeasts on the skin of apples are used and hence the apples are not washed and are collected only in dry weather. For reasons explained later (see '**The Natural Yeast Method'** page 42), like many current cider-makers, we wash off the natural yeasts, then replace them with yeast bought from a homebrew shop.

Different yeasts can produce huge differences in cider flavour, strength and colour. We deliberately use different yeasts for different barrels to bring variety of flavour to our cider.

From a homebrew shop, buy either a white sparkling wine yeast or a special cider yeast.

Do not ever use red wine yeasts nor baking yeasts!

For cider, white wine yeasts can be every bit as good as yeasts designed for cider. Some beer yeasts work very well for cider but others do not.

For your first venture, to ensure a good result, buy either a sparkling white wine yeast or a cider yeast. There might actually be a case for choosing a sparkling white wine yeast, rather than a cider yeast, because it is more likely to be tried and tested. In the recent vogue for cider-making, several new 'cider yeasts' are being marketed and not all of them are wholly successful.

Remember the seasonal rush for cider-making supplies.

There are some important things to add about specific types of yeast, including brands which can produce cider ready to drink within only three months. To avoid too much detail at once, we'll leave this till '**More about choice of yeast',** page 39.

Cider Diary notebook

Last but not least, you need a Cider Diary where you record everything you do and assess the results. That means recording the dates when you undertake each stage of this process, all your weights and measures, any observations, your tastings, and any ideas for doing things better next time. You need to do this yourself to adjust our method to your apples and your tastes. That's how we improved our cider-making, after starting with a very basic recipe.

3. STEP BY STEP GUIDE TO YOUR FIRST BARREL

Quick overview of the method

- Mash washed and chopped apples in an electric kitchen blender.
- Press this mashed apple in muslin bags in a fruit press to extract juice.
- Add a carefully chosen type of yeast to this juice.
- Prolong the resulting fermentation through adding a precise amount of sugar.
- Then store in cool, but not too cold, conditions, protected from air.
- After between three and 12 months your cider will be ready to drink.

Details will now be given for each step.

Collecting the apples

Filling a two-gallon (9 litre) barrel with apple juice is sensible for a first try. Using our ratios, that means you need 38 lbs (17 kilos) of uncut apples.

From mid-August, we keep a lidded bucket beside each apple tree and each morning collect some windfalls (leaving some for the birds, the slugs and hence the birds). Keeping grass short helps you spot windfalls. We bring bathroom scales to weigh barrels under the tree. By early September usually there's enough windfalls for pressing two or three gallons. We may pick some extra apples to make up this quantity. We generally pick apples only late in the season, when they hold most juice, and make earlier barrels from whatever windfalls are available. One consequence is that we rarely control the combination of apple types from our three different trees. If clearing windfalls comes first, you must take whatever is on the ground, though recording each mix in your Cider Diary.

Pressing the apples

Checklist for pressing: are the following available?
- At least four hours to press your first two gallons (though with practice you may need half this time).
- Fruit press, food-processor, muslin bags, lidded fermentation barrel, yeast – all as described earlier.
- Sink or trough to wash apples plus a knife for chopping them.
- Ladle to transfer mashed apple from food processor either direct to fruit press or to a storage bowl.
- Bowl for apple mashed by food processor.
- Bucket for waste residue from pressed apple and for rotten pieces.
- Lidded container so, at the end, any surplus apple juice can go to the fridge.

Do the following before you start pressing
Sterilise, or at least thoroughly clean, fruit press, food-processor, muslin bags and fermentation barrel. To sterilise, you buy 'sterilising powder' from a homebrew shop and carefully follow the instructions on the packet. These will specify a quantity of sterilising powder to add to a quantity of water and tell you how long sterilisation will take (10 minutes is typical). Don't forget to rinse off the sterilising liquid! Nowadays we use sterilising powder much less frequently than when we started. In fact, to be frank, we haven't used it for years and have encountered no ill-effects. But everything must at least be very thoroughly washed – and immediately before use. If, say, you are borrowing a long-unused fruit press or muslin bag, full sterilisation is definitely advisable. Some fruit presses are very prone to trapping moulds which could harm your cider.

Set up the press where it can easily pour into the fermentation barrel. Newspaper on the floor may be an idea here and elsewhere re splashes and spills.

Decide where to tip the bulky solid waste residue from the apples, when the waste bucket gets full. Traditionally a pig helps out here – but your compost heap or Green Waste bin may need to deputise.

Although it's the final stage, read well in advance the instructions on your yeast. Some yeasts just need sprinkling on the surface of the apple juice. But some need preparation in warm water. Often sachets of yeast are calculated for use with 5 gallons. So you'll need to measure out a proportion which matches the quantity you're making – too much yeast produces an excessively 'yeasty' flavour. Yeast quantities are so small that rarely can kitchen scales be used. Instead, we have used a tiny salt spoon to measure how many spoonfuls comprised the full packet of yeast, then calculated how many spoonfuls would measure out the required proportion. Sort this out in advance, not during the throes of pressing.

Choose a warm place where you will place the full, lidded barrel for fermentation – and be sure that you can carry it there, once it's full. (Definitely an issue for 5 gallon barrels!)

Washing and chopping the apples
Wash apples thoroughly in cold water. Chop them just small enough to fit into your electric food-processor. There's no need to peel or to remove cores. Chop out any sizeable rotten parts, but don't try to cut out every spot of brown. Throw away any apple which is rotten to the core.

Actual pressing

Fill the food-processor about a third full with apple pieces. Mash till you get a puree a bit like stewed apple.

Place the double muslin bag open in the fruit press, fill it with the mashed apple, then fold shut the bag. Screw down the press – and cloudy brown apple juice pours out into the barrel (as mentioned earlier, it will clear completely during coming months). Tipping the press will bring juice out faster. As juice flows out, you'll find you can screw the press down further. Allow time for juice to flow out.

Our Vigo press, open for loading with mashed apple, muslin bags folded over rim

Then unscrew and completely move around the contents of the bag, then screw down again and let more juice flow out. Repeat this perhaps a third time.

Then, if the bag now feels dry, throw away its contents, refill with more mashed apple and repeat the process.

Continue until your barrel is full to the top volume line marked on it. Do not fill past this lest it overflows once sugar is added and fermentation is foaming. Do not fill less than this because it's vital that there's as little air as possible between the surface of the liquid and the lid of the barrel. As will be described, air harms cider during storage.

So that we can be sure to reach the marked liquid level, we always press slightly more apples than needed – and have a jug ready to put spare juice in the fridge.

Tips about pressing
It is much, much quicker to do chopping, mashing and pressing concurrently – working the food-processor and fruit press simultaneously – than to cut all the apples first, then mash, then press them. So you screw down the press as far as possible, then go back to chopping and mashing apples, giving time for juice to trickle from the press. Then you find you can screw down the press further than before and you then return to chopping and mashing to give time for more trickling. Then you unscrew the press and move around the apple in the bag and re-press, then back to chopping and mashing.

You will need to work out for yourself how thoroughly to mash the apples, how much apple to put into the press at once, and how many times to press each bag. Different types of apple, different food processors and different presses suit different approaches. Different food processors produce different textures of apple with different degrees of juiciness.

If you mash the apple heavily, till it looks dark like overdone stewed apple, the more juice will pour into the barrel before you even start screwing down the press. But some presses do not work well on such soggy apple – it is a bit like punching blancmange. Work out how heavily to mash.

You'll need to decide which is the better way for your press – small quantities of mashed apple pressed just once or twice – or larger quantities pressed three times or even more. Once the muslin bag feels dry, no point in pressing further. If it's still wet, there's still juice in it. Though if you have lots of apples, you may not wish to spend time wringing out every last drop.

By experimenting a little, you need to work out for yourself how far to fill your food processor, how much to fill the bag in your press, and how many times to press it. When we had to replace our food processor with a different model, we found we needed to change the way we did all these things.

With our size of Vigo fruit press and the way we fill the bags, it takes around seven bags of mashed apple to press two gallons of apple juice.

Before you set up your first actual pressing for cider, you could always try a brief dummy run with your fruit press. Just assemble a few apples and get familiar with mashing them in your food processor, squeezing a bagful through your press, then drinking the resulting juice.

Fermentation
Once the barrel is full of juice, take it to a warmish place where it can stand through the initial 'fermentation' process, which you can now trigger with your yeast.

Add the yeast to the liquid, carefully following the instructions on the packet (remember to adjust the quantity of yeast powder to match the volume of apple juice as described earlier).

Place the lid so it covers the barrel closely enough to keep out the dreaded vinegar fly which can wreck your brew, but leave it loose so carbon dioxide from the fermentation can escape. For days during early fermentation, so much CO_2 will emerge that you need to have the barrel slightly open like this. But, later on, close the seal completely as soon as the airlock can handle fermentation without liquid being thrown out.

With shop-bought yeasts, fermentation should start well within 24 hours - sometimes within one hour even. Often, if you've added yeast in the evening, by next morning you'll see the first very faint traces of foam and bubbles. This bubbling then gets much much stronger and sometimes foams really furiously. This bubbling is 'fermentation', a reaction triggered by the yeast whereby the natural sugar in the apple juice turns into alcohol and releases CO_2 as a side-product.

Check that fermentation has truly got started
If it's not obvious from plentiful foaming, one test for fermentation is to listen for the quiet hissing / fizzing sound of CO_2 bubbling up. Another test is to look for the tiny bubbles of CO_2.

If fermentation seems to halt prematurely – say within a day of starting – the standard first response is to raise nearby temperature. You could temporarily place nearby a portable heater with a thermostat keeping temperature at 20 degrees C (higher than is needed by many yeasts). This has always worked for us but, if it doesn't, see page 43 for the method of last resort for restarting a fermentation.

Adding sugar to feed fermentation

Take a look twice a day and, the moment foaming seems to start subsiding, begin the next stage – feeding the fermentation with sugar. This stage usually comes between two and five days from the start of fermentation – and often around second or third day. If in doubt whether the fermentation is subsiding, err on the side of adding sugar too soon rather than too late.

If the fermentation seems undiminished for longer than five days, we would start anyway to add sugar at this point.

Alcohol is produced by natural sugar in the apple juice turning into alcohol, leaving the juice drier as a result. By adding further sugar, you are both creating more alcohol and replenishing some of the sweetness.

For your first barrel, we recommend adding sugar in the precise ratio of 28 oz sugar per gallon of juice (175 grams per litre). During your first season's cider-making, if you vary any subsequent barrels, stay within the range 24 – 32 oz sugar per gallon (150 – 200 grams per litre). In subsequent seasons you will know whether you wish to depart from this. Our recommended ratio for sugar is based on trial and error over the years and it is definitely worth following.

Weigh the total sugar you need and place it in a covered bowl beside the lidded barrel.

Use a spoon to add small quantities each day.

Very gradually add the sugar during the course of a week – not all at once!

When you add sugar, a burst of foaming or fizzing should result. During this period, you should always be able to hear and see at least a little bubbling from fermentation.

Moving to cool storage
At some point you need to move your barrel into a cool dark storage place where fermentation can continue, slowly, intermittently over coming months. If it stays too long in warmth it may end up over-strong and with an unpalatably dry taste.

The following practice has worked for us.
- After initial fermentation has subsided, one week in the same warm environment while sugar is added, as has just been described.
- Another week standing in the same warm place.
- Then move the cider to a cool dark storage place. A cellar is ideal. We don't have one, so we use a garage instead.
- Decant into demijohns with airlocks so slower fermentation can continue for months to come, protected from air. Continue storing in the cool dark place.

You have a choice about how soon you decant into demijohns.
You could do it after the cider has been settling in the barrel for a week after the last sugar has been added – i.e. at the point when you move the cider to cool storage.
Or we have sometimes left it settling in the barrel in the cool garage for 5 –7 weeks before decanting.

A gain from longer in the barrel is that much sludge will have settled by the time that you decant to demijohns (or 'racking' as it's called). Otherwise you may have to repeat the decant from demijohns, to get rid of sludge, which means losing some cider each time.

If much sludge has settled to the bottom of the barrel, it can make sense to decant through the slow method of siphoning with a plastic tube which can be kept off the bottom of the barrel. You can buy siphon tubes designed for the latter with a hole in the tube at a height above the sludge layer.

But you should delay decanting to demijohns *only if the cider in the barrel is well protected against air*, because air sours cider during storage. This means both that the barrel should be filled to the top, so that there isn't much harmful air inside and that you've installed an efficient airlock in the barrel lid. If, say, a barrel has ended up only half full, then decant into airlock demijohns *as soon as practicable* after fermentation has subsided after addition of sugar.

Likewise, if you are short of barrels and you want to make more cider, it makes sense to decant to demijohns sooner rather than later.

More often nowadays we decant into demijohns at the same time that we move the cider into cool storage – roughly a week after finishing adding sugar. During that last week we taste it once or twice to check that this period shouldn't be reduced or extended, as described very shortly. The first time you ferment cider, you may feel uncertain about what you're tasting for. So maybe just give it a week indoors anyway. But taste it a couple of times and see if you notice changes, as described shortly.

When decanting into demijohns, fill them to the top to exclude air (next topic). This sometimes means including sludgy cider which had already settled in the barrel. In due course it will settle again as the demijohn clears.

Protecting cider from air
During this period - and hereafter - you should do everything possible to minimise contact with air since this can seriously spoil your cider. While cider is fermenting strongly, we leave the lid almost, but not completely, pressed down - so CO_2 can escape. Later, when fermentation's slower, we press the lid on completely tight and CO_2 escapes through the airlock as described earlier. If you do this too soon, you may be flooded with foam coming up through the airlock.

Fullest protection against air comes from demijohns with water-filled airlocks. You should fill them up with cider very close to the cork to minimise air. Check periodically whether the water in the airlocks needs topping up because of evaporation. Extreme cold can freeze airlocks, as discussed on page 29.

Keep the following in mind
Cooler surroundings slow down or even halt fermentation. If initial fermentation has been weak, brief or erratic, you could prolong the time in warmth and delay decanting so as not to halt fermentation prematurely. Likewise if the fermenting liquid still tastes very sweet, which means that there's much room for more conversion of sugar to alcohol.

Conversely, if the fermenting juice is already starting to taste noticeably drier, it is time to move somewhere cooler and to decant sooner rather than later.

If you leave behind all the yeasty sludge when you decant (called 'racking'), you may much reduce or halt further fermentation. If you stir up that sludge instead, you may revive a weak fermentation. If the juice already tastes dry, the former might be helpful. If it's still very sweet, then the latter.

If the weather is especially cold, it could be wise to delay before moving the cider to outbuildings – severe cold will be discussed shortly.

Some yeasts need higher temperatures than others to sustain fermentation. Some yeast packets carry thorough instructions on the necessary temperature range. You could use more delicate yeasts in September, rather than November. See '**More about choice of yeast**' on page 39.

Close-up of airlock in fermentation barrel lid, secured by a washer. Lower water level on left indicates fermentation.

You can tell that fermentation is continuing if there is a difference between the two water levels in the type of airlock in the photo above. The level on the left is lower. (While this photo shows the lid of a fermentation barrel, not a demijohn, the same thing applies with demijohns.)

If fermentation is vigorous, straightaway you'll see bubbles of CO2 passing through the water. When fermentation is still alive but very slow, maybe bubbles won't be obvious but the water on each side of the bend in the airlock will not be level. See how in the photo the water level on your left is slightly lower. The CO2 from fermentation is pushing the water upwards round the bend and, if watched long enough, there'll be a bubble.

This difference in water levels shows you at once that slow fermentation is continuing. When the airlock's water is level on both sides of the bend, that means fermentation has halted.

Decanting into bottles
You need *both* the following signals that it is time to do this:

When fermentation has halted according to the signs just mentioned. You can't always get this right and fermentation may revive and this needn't be a problem. But never bottle cider which is still strongly fermenting or it will blow the cork off!

When sediment has settled and the cider is clear. (Our cider almost always clears with passage of time but sometimes it has been necessary to clear cloudiness with 'finings' as described later. Sometimes cloudiness can seriously impair taste – and this can be reversed by finings.)

To decant into bottles, we siphon from the demijohn so as not to disturb the sediment on the bottom. We siphon into a large jug with a fine spout. From this jug we can fill bottles precisely to close to the top so that little space is left for the air which can sour cider. There's less cider spilled than if you siphon directly into bottles.

Protecting cider from air (again!)

We keep our demijohns outside in the dark, cool garage all summer. When we open a demijohn, we decant it completely into well-filled, corked bottles, which are then kept in the darkest, coolest part of the garage. This avoids 'air souring' within a half-full demijohn.

For the latter reason, some people buy winebox-like containers, like Vigo's Manucube, which dispense cider without air contact.

When we bottle cider, quite often we use bottles which are only half litre size or even smaller. This is to avoid leaving cider exposed to air in a half-drunk bottle, when one does not wish to drink a full 75 cl bottle.

When will the cider be ready to drink?
'When it tastes ready' is the answer you'll get from many cider-makers. It depends on your apples, your yeast, your storage process, the weather, and your tastes – and it may vary from year to year. The milder the winter, for instance, the more that fermentation continues in cellar-stored cider, so the earlier your cider will be ready.

Between five and nine months from pressing is a common period, though some people like to leave it a full year.

By way of contrasts, in one recent year a barrel from one of our yeasts became good to drink after only three months, one from a different yeast after six months, while a third yeast needed a full ten months. Often our cider has taken ten months to reach its best. Recently we had very dry ciders from two different yeasts which took a full 12 months to become pleasant to drink. (More about 'early drinking' yeasts in '**More about choice of yeast'**, page 39.)

Even if it's ready early, hold some of your cider back to see how good it can get if left longer. Around May, the rising temperature often triggers another round of fermentation in cool-stored cider so it's worth checking how your cider tastes after this. Hence the old English country adage about cider not being ready till the first call of the cuckoo.

But, in our experience, leaving cider longer doesn't always lead to improvement. Some of our best cider has actually become too dry and too strong for our tastes by 18 months or two years after pressing. This has happened with our favourite early drinking ciders from Vintner's Harvest MA33 yeast and Young's Cider Yeast, which seem best between 3 months and 12 months after pressing (see '**More about choice of yeast',** page 39). Almost certainly this change in taste is because we don't use Campden Tablets to artificially halt fermentation – see Campden Tablets on page 37 for explanation. We still won't use Campden Tablets but will now drink all the cider from these yeasts within a year.

4. TROUBLESHOOTING

During 16 years of making cider we've encountered problems only rarely. But here's a list of possible problems – and what you can do about them.

Please note that there's a case for leaving this section of the book until after you have got your cider into demijohns – and move on to Chapter 5 until then. The reason is that it may feel confusing to dwell on problems which can't arise until long after fermentation and which you may never encounter anyway. It can make cider-making seem harder than it is.

If you've bought this book well before your apple harvest and you are itching to get prepared, there's a more helpful next step than reading this troubleshooting section prematurely. You could just buy a few apples and practice making juice with your food processor, fruit press and muslin bags.

Extreme cold weather
Spells of very cold weather in early winter can halt fermentation altogether and this may spoil the cider.

During England's unusual extremely cold winter of 2010, we responded by moving all our demijohns of cider back from the cold garage. First we briefly placed them in a warm room to revive fermentation, which had halted completely. In a demijohn you can spot continuing fermentation through the bubbles coming through the airlock or tiny bubbles in the liquid. Then, from December till the end of March, we kept them in a semi-heated conservatory with temperature in the range 8 – 10 degrees Centigrade. Then back to the cool dark garage. Results were good.

This course of action is safest but may not always be necessary. In traditional English cider storage, fermentation slows to the imperceptible during freezing January / February weather but later revives without any intervention. However the unusually early extreme freeze of early December 2010 is now said to have caused harm to some cider. Maybe a strong freeze can harm yeast at that earlier stage. We're not going to risk this!

If you don't move your cider somewhere warmer, beware lest the water in the airlocks freeze for then they won't work. Some people add vodka to the water in the airlock in order to stop it freezing.

Cider which tastes or looks bad
During our many years of home cider making we must have made well over 50 separate barrels of cider. But never once have we had to throw away a barrel.

Several times we've produced a barrel which tasted or looked so horrible that we were sure it would have to be thrown away. But, one way or another, they all ended up perfectly drinkable.

By the time 12 months has passed, if your cider still isn't pleasant to drink, it's high time for intervention. Here are a few tools for modifying poor-tasting cider:
- Blending with other cider or with shop-bought apple-juice.
- Adding a second lot of sugar.
- Mixing it with spices and serving as warm, mulled cider.
- Adding 'finings' – a substance which makes the muck in cloudy cider settle as sludge on the bottom.

- 'Racking' – decanting the liquid to get rid of the sludge on the bottom.
- Last but not least, giving it yet more time to mature.

Blending problem cider with other drinks

We've had excellent results from blending over-sweet and over-dry cider. Two undrinkable ciders can make a very tasty marriage! We always hang on to our occasional unpalatable demijohns, waiting for a suitable partner for blending, so two wrongs can make a right. It may just be imagination, but these blends seem to taste much better a few weeks after mixing than immediately afterwards.

Usually it takes only a little over-sweet cider to improve over-dry cider. Add sweet to dry and check the taste. Adding 1 part sweet to 3 or 4 parts dry may do the trick.

We've also had good results from blending over-dry, over-strong cider with shop-bought apple juice.

Adding a slice of orange can reduce sweetness, though not to everyone's taste.

Resugaring unpleasantly dry cider

In our early years we quite often produced cider which was unpleasantly on the dry side – possibly as a result of over-fast fermentation (too long in warm conditions) or possibly through air-souring.

A simple solution proved to be adding a little extra sugar to the mature cider, something which we're finding that quite a lot of cider-makers do. We would do this in a ratio of 3 oz extra sugar to 1 gallon of cider.

In metric, this is better phrased as 19 grams of sugar per litre, since it's not wise to resugar a whole gallon unless it will be drunk fairly soon. To help sugar dissolve, you could use caster sugar or first stir it into a smaller amount of cider.

Some key points:

- Resugaring may trigger secondary fermentation, so it should be done in a container with an airlock. If not, the top may get blown off. You may get pleasantly sparkling cider.

- It usually takes between *one and three weeks* for a major improvement in taste to occur. Immediately after the extra sugar is added it may taste like, well, cider with sugar tipped in. But wait a little and then poor-tasting cider gradually turns delicious – though only for a while.

- In our experience, this improvement in taste often proves short-lived. Quite soon the cider may revert to the original over-dry taste, though stronger now due to secondary fermentation. So don't resugar more cider than you'll drink before the improvement starts fading.

Resugaring can temporarily turn around a batch of cider which you otherwise wouldn't want to drink. But it is also a nuisance, because of the need for timing, just described. We've only rarely needed to resweeten any cider for some years now and this is so much better. Nevertheless, some years it has still been necessary.

If all your cider needs resweetening, it's probably a sign that your procedure could be improved.

Resweetening with honey, not sugar

Some pleasing variety can come from using honey for the above purpose, instead of sugar. Sometimes, though not always, it adds something really special to the taste.

- Try runny honey in a ratio of one part honey (by volume) to twenty parts cider.

- Take care to stir the honey in really well or it may just collect on the bottom. Maybe pour out some cider from a demijohn, stir in honey vigourously and shake, then return to the demijohn.

- Expect the same time sequence as for sugar.

Warm, spiced cider to improve taste

You can sometimes mask an unsatisfactory taste by making warm spiced cider. There are loads of mulled cider recipes on the internet (search **www.cooks.com** for instance). Here is a quick recipe which we used to improve an oversweet, slightly vinegar-flavoured cider from a tricky natural yeast fermentation.

To serve two people

1. To one mug of orange juice add 1 teaspoon powdered cinnamon, half teaspoon powdered allspice and quarter teaspoon powdered cloves.
2. Bring to boil in a pan, then simmer for 15 minutes.
3. Add one mug of cider to the pan, stir well and briefly warm but don't boil.
4. Pour through tea strainer into mugs, adding a little vodka to each.

Rationale behind this recipe:

- Boil the spices in orange juice, not cider, to avoid evaporating the alcohol in the cider.

- The reason for using powdered spices was to make it quickly. You'll get a much nicer looking, less cloudy drink by using whole cinnamon, cloves etc, but you'd need to leave these spices in the drink for much longer.

- Orange juice was used because this was an oversweet cider. Had it been over dry, mix with apple juice instead.

- With medium cider, the orange juice may give too tart a taste. Better to use apple juice instead.

Clearing cloudy cider
Cider produced through our method starts off cloudy, as already mentioned, but almost always clears completely.

Occasionally some cider stays cloudy even after time to settle in a demijohn, while other demijohns have cleared completely.

To clear cloudiness, you can mix in a substance called 'finings', bought from a Home Brew Shop. The 'finings' cause the cloudiness to settle at the bottom of the demijohn. Then you decant the clear liquid. With severely cloudy cider, finings may greatly improve the taste as well as the appearance.

There are many different brands of finings and sometimes we've found it hard to obtain the same brand the following year.

Faced with a choice of finings for wine or for beer, we've found it was wine finings which suited our cider.

Sometimes finings clear cider completely within 24 hours. Sometimes nothing happens, then weeks later it suddenly clears and you are unsure whether the finings played a part. Sometimes two lots of finings are needed. Sometimes no improvement ever occurs. Sometimes the same finings work wonders on one demijohn but leave another unchanged.

Our preferred brand of finings

A British brand of finings, which we've particularly liked, is Young's Wine Finings (see Chapter 6). It involves a complicated, two-part process but gave us excellent results. Young's Wine Finings has completely cleared some demijohns of very muddy cider on which 'Beer Finings' and 'All Purpose Finings' had little effect. Young's Wine Finings involves adding two different fluids to the cider in sequence, as instructed on the packet.

- A 10 ml syringe with a millilitre scale proved useful for measuring the right amount of each component of Young's Wine Finings to fit the volume of cider for which they're needed.

- A barbecue skewer is useful for stirring the finings inside the demijohn, since it's thin enough to get inside.

- Clearing can be expected after 18 hours – but best leave another day before decanting.

- Through the action of these finings, expect much residue to sink to the bottom of the demijohn and sometimes scum floating upwards.

- Once you've thus cleared some cider, best decant it soon and get rid of all that residue deposited by the finings.

Cloudy cider, which looks like a muddy puddle, can quickly become lustrously clear through adding Young's Wine Finings.

Mouldy cider
We've only ever encountered this from the natural yeast method. There's not much to do except keep removing the mould. And avoid the natural yeast method!

Steering cider towards sweeter or drier

At fermentation stage
To recap on what's already been said, the stronger and longer the fermentation, the more you'll get dry (and strong) cider. Length of time in warmth prolongs and strengthens fermentation.

Sweeter, weaker cider results from:
- shorter or weaker fermentation:
- early transfer to cool conditions
- early decanting of the cider into demijohns, leaving the yeasty sludge behind (called 'racking'), can halt fermentation.

But there isn't a straightforward link between amount of sugar added soon after initial fermentation and eventual sweetness. You can add lots of sugar but end up with very strong but dry cider, if fermentation has been so strong that the extra sugar has been converted to alcohol.

Early transfer to coolness has been the tool which we've found important for ensuring sweeter cider. Combine this with our upper ratio for adding sugar – and a sweet cider will be likely indeed.

A drier cider is likely from delaying transfer to cool and delaying transfer to demijohns. Cider's natural trend is to become gradually drier.

It's during your second cider harvest that you can try to steer your cider's taste, because it's not till then that you'll know how you want to change it. Though there are no certainties about results – and the surprises from each year's new tastes are part of the fun of cider making.

Post-fermentation: limiting the drift to dryness

Campden Tablets
A tool for halting the drift to dryness is Campden Tablets (potassium metabisulphite), which many home cider makers use. You but them from a Homebrew Shop. These tablets halt fermentation artificially from the moment they're added and hence halt the steady process whereby cider gradually gets drier.

We've rarely ever used Campden Tablets, disliking chemicals. But cider-makers should know about this option.

To use Campden Tablets, you need to decide when to halt fermentation – either on the basis of taste or, more advanced, by measuring the percentage of alcohol. And the latter is too advanced for us!

Some cider makers also routinely use Campden Tablets for a separate purpose. As well as washing the apples, they use Campden Tablets as a pre-fermentation yeast-killing treatment to stop wild yeasts interfering with a shop-bought yeast, which they subsequently introduce.

For this purpose, we've only ever used washing the apples and we've never encountered problems when using shop-bought yeasts. You can definitely make excellent cider without ever using Campden Tablets, either before or after fermentation.

However, let's acknowledge a downside to not using Campden Tablets for the purpose first mentioned, halting progress towards dry and strong. Sometimes continued fermentation can be a nuisance. You may bring a bottle indoors and the increased temperature restarts fermentation so the top pops off the bottle. If fermentation continues, your cider will keep changing taste, getting gradually drier and stronger and this may not suit your tastes.

'Racking'
The non-chemical way to halt continuing fermentation is 'racking'. This means decanting the cider to another demijohn so that you leave behind the sludge on the bottom, where residual yeast is lurking. This certainly doesn't always work. And you'll lose some cider each time you do this.

Post-fermentation: encouraging the drift to dryness
Sometimes we've had cider which, more than a year after first fermentation, still remained somewhat sweeter than we wanted. We've sometimes managed to make it slightly drier by bringing the demijohn in from the cold winter garage to a warm room. There some fermentation revived, as shown by uneven water levels in the airlock on the demijohn and tiny bubbles as described earlier. After a month or two in a warmer place the taste became drier as desired.

5. MISCELLANEOUS

More about choice of yeast

We started paying much attention to choice of yeast after reading an interesting internet article by Greg Appleyard, a home cider maker in Canada. He had organised a careful experiment whereby a panel of friends judged homemade ciders which had all been made from the same apple juice but with many different yeasts. They rated them on taste, smell and colour - and this experiment showed all of these to be strongly affected by the type of yeast. Best ratings for taste went to sparkling white wine yeasts, cider yeasts, and certain of the natural yeasts. Red wine yeast was the type consistently rated worst. Some beer yeasts were rated well and others were rated badly. Different natural yeast ciders got either very good or extremely bad ratings (see later).

This experiment got us carefully comparing results from different types of yeast and we've found that yeast has an enormous effect on taste and strength. We now deliberately use different yeasts for different barrels to bring pleasing variety to our cider.

Our current favourite yeasts
In recent years we've been delighted by an elaborate range of yeasts from an international firm called Vintner's Harvest (see page 47), which are intended to suit wines made from different fruits. Their yeasts have given luscious, contrasting tastes to our cider.

Particularly noteworthy, their MA33 yeast has given us cider which can be delicious to drink after only three months. This contrasts dramatically with the 9 – 11 month maturation periods to which we'd previously become accustomed. Sometimes we can ferment a barrel with Vintner's Harvest MA33 in early September and by December it's lovely to drink and can be bottled for Christmas presents.

Of those other yeasts, which Vintner's Harvest recommend for apple wine, we've had good, subtle-tasting cider from Vintner's Harvest CY17. This however takes a long while to mature – at least nine months and it's better after 12 months or longer. (In fact we've now tasted it 30 months after fermentation and the extra time has wonderfully deepened its luscious taste.) CY17 also needs quite high temperature. It's not the best choice for your very first barrel.

Vintner's Harvest SN9, by contrast, ferments easily and at low temperatures. Sometimes SN9 can be good to drink after six months, though sometimes it needs much longer. However, we've had variable results with this one.

In Britain a promising yeast brand is Youngs Cider Yeast. It's not dramatically superior to everything else, just because it is made with cider in mind. But it is good, can add a pleasant, moderate bite to the taste, and we regularly use it. Sometimes the results have proved excellent to drink after only three months – but sometimes this yeast has needed many months longer. In contrast to CY17, we've preferred this cider when it's quite young: some vintage bottles have become too strong.

If seeking an early drinking cider for your first barrel, compared to Vintner's Harvest MA33, Young's Cider Yeast might be the safer bet since the former needs quite warm temperatures. Best to minimise complications for your first barrel.

If you can't get the specific yeasts mentioned here, for your first barrel seek either any sparkling white wine yeast or a cider yeast – but nothing else. Sparkling white wine yeast is a safer bet than some newly marketed cider yeasts. We have recently tried some of these new cider yeasts and found they produced cider which was much too dry and lacking in flavour for our tastes. Of those cider yeasts which we have tried, Young's Cider Yeast is the only one which we would use again. Best perhaps not to start off with a cider yeast unless it's a brand for which you have received a specific recommendation from someone who has tasted the end product. However, the best-laid plans of cider makers can unaccountably go astray and, for all we've learned about the usual merits of the yeasts recommended here, occasionally they've performed badly.

There's a good sparkling white wine yeast which you can easily get by mail order. This is W0094 sparkling white wine yeast which the Devon cider specialist **Vigo** will post you as a single packet (enough for 6 gallons) in an envelope at ordinary postage cost. This cider can be drinkable after only three months, if on the sweet and weak side. But by 11 months after fermentation it can change markedly to become stronger and much more tangy – more of a classic English cider taste. Don't drink it all before it reaches this stage!

Using natural yeasts for fermentation

A surprising finding from Greg Appleyard's yeast experiment was that the very best rated tastes of all came from natural yeasts, the yeasts which grow naturally on apples on the tree. While traditional English cider-making always used these natural yeasts, nowadays home cider makers generally avoid them because of their reputation for causing mould. (Indeed, in Greg Appleyard's experiment some natural yeast fermentations also produced the worst-rated ciders.)

So Greg Appleyard's experiment started us trying natural yeasts in a quest for a super-marvellous taste. Sadly it hasn't worked out for us and, after some years of trying, we've now dropped the method. Only once have we obtained an exceptional, superbly delicious taste, different from shop-bought yeasts. But this came from an unusual low sugar, natural yeast brew which grew such horrendous white mould that we would not repeat it.

We've produced many other good tasting ciders from natural yeast. But they've been no more than equal, not superior to, shop-bought yeasts. Set this against the trickier aspects of natural yeast – the greater risk of weak or hesitant fermentations and of cloudiness, moulds or strange tastes – and on balance we've decided to stick to shop-bought yeasts. Still, if you are curious, this is how to ferment with natural yeast.

Practical application of natural yeast fermentation

To preserve the yeasts on the apple skin, you must gather your apples after a spell of dry weather and must not wash them before chopping. (So here come those moulds!) Everything else during pressing is exactly the same as already described – except of course that you don't add a packet of yeast!

It's especially important that the barrel goes in a warm place because natural yeast fermentation can be much harder to start.

Expect it to take 48 hours for a natural yeast fermentation to start – much longer than with shop-bought yeasts.
Be ready to intervene if fermentation fails to start or falters.

Remedying failed or stuck natural yeast fermentation
If fermentation doesn't start, your laboriously pressed juice may spoil. Two tacks we've employed with problem natural yeast fermentations are first extra warmth, then extra natural yeast from adding apple-skin peelings or letting whole apples bob around on the liquid. It can be a fraught business, waiting for fermentation with heaters surrounding the barrel and apples floating inside it. In our experience natural yeast fermentations have always eventually started, though we've never known which, if any, of our interventions were assisting this.

What do you do if even this fails to start fermentation? The next step could be to decant a glassful of the apple juice and, as described next, add a very little shop-bought yeast to get this glassful fermenting, then return it to the barrel to act as a trigger. Only once have we ever begun this step – but, by the time the glassful was fermenting, natural yeast fermentation in the barrel had begun anyway.

Remedying failed or stuck fermentation with shop-bought yeasts
We've never ever had any fermentation problems with shop-bought yeasts, which didn't get solved simply by increasing warmth. But, if warmth didn't work, the last resort would be to decant a glass of the juice and add a very little shop-bought yeast, stir it well and keep it warm and wait till this glassful was itself fermenting.

Then pour it back into the barrel, stirring well, for it to act as a trigger. If you didn't have any of the original yeast left, note that Vintners Harvest recommend their SN9 yeast for thus restarting failed fermentations by other yeasts. But expect that you'll never encounter this problem. The only time we've ever actually used this trigger method was with a feeble fermentation of blackberry wine – and it did the trick very swiftly.

Occasionally we've had a successful fermentation which then has seemed to finish unusually quickly, say during the week after finishing adding sugar. This has occurred with yeasts which need a slightly higher temperature, like Vintners Harvest MA33. We have then used a blow heater with a thermostat to raise the temperature to around 20 degrees Centigrade. How long should you keep the temperature raised this way? We kept tasting the fermenting juice every couple of days. In less than a week it came to taste as dry as we'd want it. So the heater was turned off, normal procedure resumed, and nature was left to take its course.

Saving money on the fruit press?
If your garden produces enough apples to make cider, over the years the costs of a fruit press will surely pay off. But, for anyone who wants to save on buying one, the following options exist. However, we've never tried either option ourselves.

Not using a press at all – a wasteful option
During our process, when you fill the muslin bag in the press with mashed apple, you'll notice that much juice pours out before you even start screwing down the press. Thus it seems likely that you could extract a fair amount of juice without a press as follows.

Intensely mash the chopped apple in the blender till a dark sloppy mess. Then put it into a muslin bag. Then place the bag in a colander over a bowl and press and squeeze by hand, before emptying the bowl into your fermentation barrel.

It would be a wasteful process, which would miss a lot of juice. But you still could collect a considerable amount.

Making your own fruit press
On the internet are many instructions for making various types of apple press. On You Tube you can watch folk showing off their home-made cider presses in action – including some ingenious, non-traditional models, using car jacks and the like. We haven't actually tried any of these, so our comment is limited. But ask yourself the following, if thinking of making your own press.

Can you make a press which is strong enough to last?

Could you quickly replace broken parts? There is massive mechanical strain in pressing just our garden's volume of apples. We changed to a steel Vigo press because we repeatedly cracked the wooden pressure block in the traditional wooden fruit press with which we first made our cider. Each apple harvest we had to saw a new wooden pressure block to have on standby. One certainly wouldn't want to be suddenly stuck with a broken press, half-full juice barrel and barrow loads of apples.

In You Tube home-made cider press videos, you can't tell whether you're watching a home-made machine which has run successfully for years – or its first five minutes of operation. Sometimes the pressing in these videos is occurring outdoors - something which people would only do on their first attempt, considering how many insects want to join in.

How fast can these home-made presses work, compared with a shop-bought press? Sometimes in these home-made cider press videos the liquid containers are tiny. Could this press handle the volume which you might want to process? A 5 gallon barrel is hard work even with a good press.

Note that powered and multi-layer presses are among the models presented. You could be better off attempting the much simpler model described here, the 'cylinder wherein a block is screwed down by hand'.

Watch out for a single key component, like a special screw, whose high cost may make DIY a false economy.

Whilst giving these cautions, we have also seen large home-made presses which were reliable, which massively cut the time needed for pressing because they were so big, and which cost less than commercial equivalents. Hats off to people with the skills to make these!

6. CIDER-MAKING SUPPLIES AND INTERNET CIDER RESOURCES

Cider Making Supplies
Vigo Ltd
Vigo Ltd is a major supplier of every sort of hardware you might need – and they also sell yeasts and related substances. See their website for their lists of products. They're based in England's cider heartland.

Address: Vigo Limited, Dunkeswell, Honiton, Devon EX14 4LF.
Telephone:+44 (0) 1404 892101
Email: **sales@vigoltd.com**
Website: **www.vigopresses.co.uk**

Vintners Harvest
At **www.vintnersharvest.com**, you can read about the varied and intriguing range of wine yeasts developed under this brand – some of which we have used repeatedly for varying the taste of our cider. Vintners Harvest MA33 has become our particular favourite.

Vintners Harvest is an international brand, probably based in New Zealand, which seems on sale in many places. Not sure how you'd actually order yeast directly from them, for their website is as clear as a bad barrel of natural yeast cider as far as mail order and company location are concerned. We've got their yeasts from a nearby small home-brew shop. If you Google, you may find you can buy their yeasts from many sources not listed on their own website.

Young's Home Brew
In Britain, Young's supply a special Cider Yeast and effective finings.

Address: Young's Home Brew, Cross St, Bradley, Bilston, West Midlands W14 8DL, UK
Telephone: +44 (0) 1902-353352
Email: **sales@youngsgroup.co.uk**
Website: **www.youngshomebrew.co.uk/**

Suppliers in North America
See under 'Press Suppliers' in **www.cider.org.uk**

Websites to further your knowledge
This booklet contains all we needed to know to produce our 10 – 20 gallon annual cider supply. But if you wish deeper knowledge, try the excellent website, chat forums and links lists of Andrew Lea (**www.cider.org.uk**). Andrew is a former professional English cider maker, now an exceptionally knowledgeable, scientific, garden-scale cider maker. His website is vast and is well attuned to home cider makers, large-scale or small. Anything you could want is there – suppliers of particular apple trees or different types of fruit press, for instance, or cider making science and history.

Varieties of apple
An aspect of cider-making is using particular varieties of apple tree for the taste they give to cider or planting them for this purpose. We have never explored this, since we focus on the apple trees already in our garden. But, when occasionally we have pressed someone else's apples, it's amazing how different apples can be. If you can pick and choose from apples in other people's gardens, you could design your apple blends carefully. Andrew Lea's website (**www.cider.org.uk**) has a section about choosing apple trees.

Contact with other cider-makers

On Andrew Lea's list, there's a British participative web chat group, Cider Workshop (**www.ciderworkshop.com**) and a North American network, Cider Digest (**www.talisman.com/cider/#Digest**). Anyone can join either network - the British network has some Canadian contributors, for instance. These are valuable opportunities to raise queries with experienced cider makers.

How simple do you want your cider making?

Alas, at some point or other, all these resources go over our heads! We've encountered quite a few books about 'simple home cider making' - but none which was quite simple enough for us. So we thought we'd write it ourselves. Hence not a word about measuring 'specific gravity'!

There's a whole world of extra information out there for home cider makers who want to get into different apple varieties, measuring and manipulating alcohol strength and more advanced techniques. Certainly much extra finesse is possible and it will certainly bring benefits. It's for you to choose. But we've found that you can still make much very pleasant cider without knowing these things.

Yes, this booklet takes a very amateur approach and seeks to serve complete beginners. But even some experienced DIY cider makers might gain from certain things which we've learned:

- Type of yeast has enormous influence on how cider tastes. If you haven't yet focused on yeast selection, give this a try.

- Try our magic ratio of 1.75 lb sugar per gallon, as initial fermentation declines.

- If you don't do it already, focus on cool, dark, air-tight storage of cider. This so much improved our cider

- Don't throw away poor cider – it's amazing how again and again dismal cider can turn out fine, given time or blending. Usually it's within 12 months but, after waiting 2 years, even our weirdest-tasting natural yeast cider turned into a pleasant luscious drink.

6.   ICING ON THE CAKE

There are various amusing or interesting things you can do with cider, which don't need to be detailed in this summary of essential instructions.

Cider recipes

On the internet you can find for yourself masses of recipes for spiced cider, chilled cider, cider shandy, carbonated cider, cider cocktails, and so on. And also for cooking with cider, if this catches your fancy.

Fun with labels on your bottles

If you've made a variety of ciders, via the yeast and sugar variations which we've outlined, you could give each a name and maybe design a label for your bottles. What with internet clip art and Photoshop or decoupage and scanner, there's plenty you can do.

Places to visit

Here and there are some museums about the traditional cider industry. Particularly interesting are traditional cider works which are still operating, where you can see the equipment, ask questions, and try the product. In Britain's cider heartlands, like Somerset, Devon and Herefordshire, there are very many cider makers. For a comprehensive list, county by county, see:

www.ukcider.co.uk/wiki/index.php/Cider_and_Perry_pr oducers_-_full_list_arranged_by_county

Even in counties which we don't normally associate with cider, you'll find somewhere to visit on this list.

A visit which we particularly enjoyed was to Brimblecombe Cider near Dunsford, Devon on the above website's Devon list. Splendidly traditional equipment and excellent vintage cider to sell, plus a pleasing touch of the anarchic attitude of some traditional Devon people. Visit and you'll understand!

Andrew Lea's website has two much shorter lists showing cider heritage resources and these include US and Canada and continental Europe. Scroll down **www.cider.org.uk/frameset.htm** and also view its 'Further Resources' link.

Then there are the various cider festivals. This website has sometimes displayed a list, though not regularly updated: **www.ukcider.co.uk/wiki/index.php/Current_events**

Happy cider making!
Yes, there are many enjoyable strands to cider making. And perhaps the best is experimenting and customising cider to your own tastes.

Enjoy your cider making! And part of truly happy cider making, let's not forget, is not drowning yourself in the lake of free alcohol which your apple trees can produce.

Lastly, here's a first entry, which you could put in your Cider Diary. Please write a reminder, when your cider is ready, to review this book on the Amazon website. Reviews are slow to come, which is only natural, for how can you really review this book till you've tasted the cider which it's helped you to make? Please don't forget, because readers' reviews are crucial for showing whether this book really delivers the goods.

The End

Index of Topics

Fitting an airlock into the lid of a plastic fermentation barrel

First you make the hole. Then fit the rubber washer into it. Then insert the plastic airlock. Photo of result on page 25.

Making the hole

The rubber washers, sold for this purpose at homebrew shops, are designed for a specified size of hole. Some UK fermentation barrels have a circle already marked on the lid to guide you.

For a common UK washer, you need to make a hole with a diameter between 7/16th and 9/16th inch. Airlocks can show fine variations in their thickness. Start at the smaller size for the necessary tight fit.

One way to make the hole is with a large drill bit. You can enlarge the hole as needed with the sides of the drill bit. You could also make the hole just with a gimlet and slowly enlarge with a circular rasp or something similar.

Fitting the washer

Squeeze the washer into the hole. Then, working on alternate sides of the lid, get the washer's lips to grasp both sides of the hole. It's not difficult.

Inserting the airlock

Make sure it's not too deep, for you need to fill the barrel very full to minimise air contact. The airlock must not dip into the juice or CO_2 won't be able to get out via the airlock. If the airlock is too loose in the washer, you could simply wrap sellotape round the lower end of the airlock tube, so it fits tighter. Or you may need to seal it with sealant or glue – on the upper side only! But this usually isn't necessary.

ABOUT THE AUTHOR

'Charlie Henley' is a pen-name. However, the author is in fact related to the historic Henley dynasty of cider-manufacturers in the beautiful village of Abbotskerswell in Devon, south-west England. Though he never discovered this till well after making his first hundred gallons of cider.

By the same author
How did Charlie fill his autumn evenings before he moved to a home whose garden offered so much cider-making opportunity? See the booklet listed next.

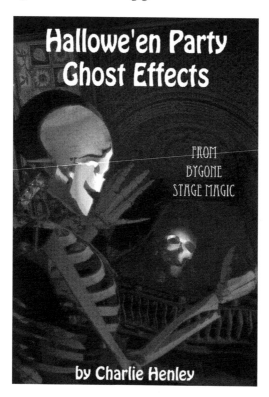

Printed in Great Britain
by Amazon